THE NIGHT SKY and Other Amazing Sights in Space

Eclipses

Nick Hunter

Chicago, Illinois

an imprint of Capstone Global Library, LLC
Chicago, Illinois

To contact Capstone Global Library please phone 800-747-4992, or visit our website www.capstonepub.com

Edited by Rebecca Rissman, Daniel Nunn, and Sian Smith
Designed by Joanna Hinton-Malivoire and Marcus Bell
Picture research by Mica Brancic
Production by Sophia Argyris
Originated by Capstone Global Library Ltd
Printed in the United States of America in North Mankato, Minnesota. 102013 007857R

17 16 15 14 13
10 9 8 7 6 5 4 3 2 1

Library of Congress Cataloging-in-Publication Data
Hunter, Nick.
Eclipses / Nick Hunter.—1st ed.
p. cm.—(The night sky: and other amazing sights)
Includes bibliographical references and index.
ISBN 978-1-4329-7515-9 (hb)
ISBN 978-1-4329-7520-3 (pb)
1. Solar eclipses—Juvenile literature. 2. Lunar eclipses—Juvenile literature. I. Title.
QB541.5.H86 2014
523.7'8—dc23 2012043047

Acknowledgments

The author and publisher are grateful to the following for permission to reproduce copyright material: Alamy p.12 (© Images of Africa Photobank/David Keith Jones); Capstone Publishers pp.28, 29 (© Karon Dubke); Getty Images pp.4 (Photolibrary/Steven Puetzer), 10 (AFP Photo), 11 (AFP Photo/Damien Meyer), 20 (Science Faction/Jay M. Pasachoff), 25 (De Agostini Picture Library/G. Dagli Orti); NASA pp.15 (JPL-Caltech/R. Hurt (SSC)), 23 (Hinode/XRT); Science Photo Library pp.8 (Gary Hincks), 13, 14 (Babak Tafreshi, Twan), 18, 19 (John Chumack), 21 (Laurent Laveder), 22 (ESA/STSCI/E Karkoschka, U. Arizona), 24 (John R Foster), 26 (Mark Garlick), 27 (David Parker); Shutterstock pp.5 (© Albie Bredenhann), 6 (© Kokhanchikov), 7 (© sdecoret), 9 (© Viktar Malyshchyts), 16 (© Sergey Khachatryan), 17 (© Mark Bridger).

Cover photograph of a solar eclipse reproduced with permission of Shutterstock (© Igor Kovalchuk).

We would like to thank Stuart Atkinson for his invaluable help in the preparation of this book.

Every effort has been made to contact copyright holders of any material reproduced in this book. Any omissions will be rectified in subsequent printings if notice is given to the publisher.

Contents

Out of This World 4
Day and Night 6
Earth, Sun, and Moon 8
Solar Eclipse 10
Total Eclipse 12
How Do Solar Eclipses Happen? 14
Animals and Eclipses 16
Eclipse of the Moon 18
Exploring Eclipses 20
Across the Solar System 22
Eclipses in the Past 24
See for Yourself 26
Model an Eclipse 28
Glossary *30*
Find Out More *31*
Index *32*

Some words are shown in bold, **like this**. You can find them in the glossary on page 30.

Out of This World

Look up into the sky on a clear night and you will see hundreds of lights. These are the **stars** and **planets**. They share the **universe** with our own planet, Earth.

You can look for stars and planets in the night sky. Stars twinkle, but planets do not.

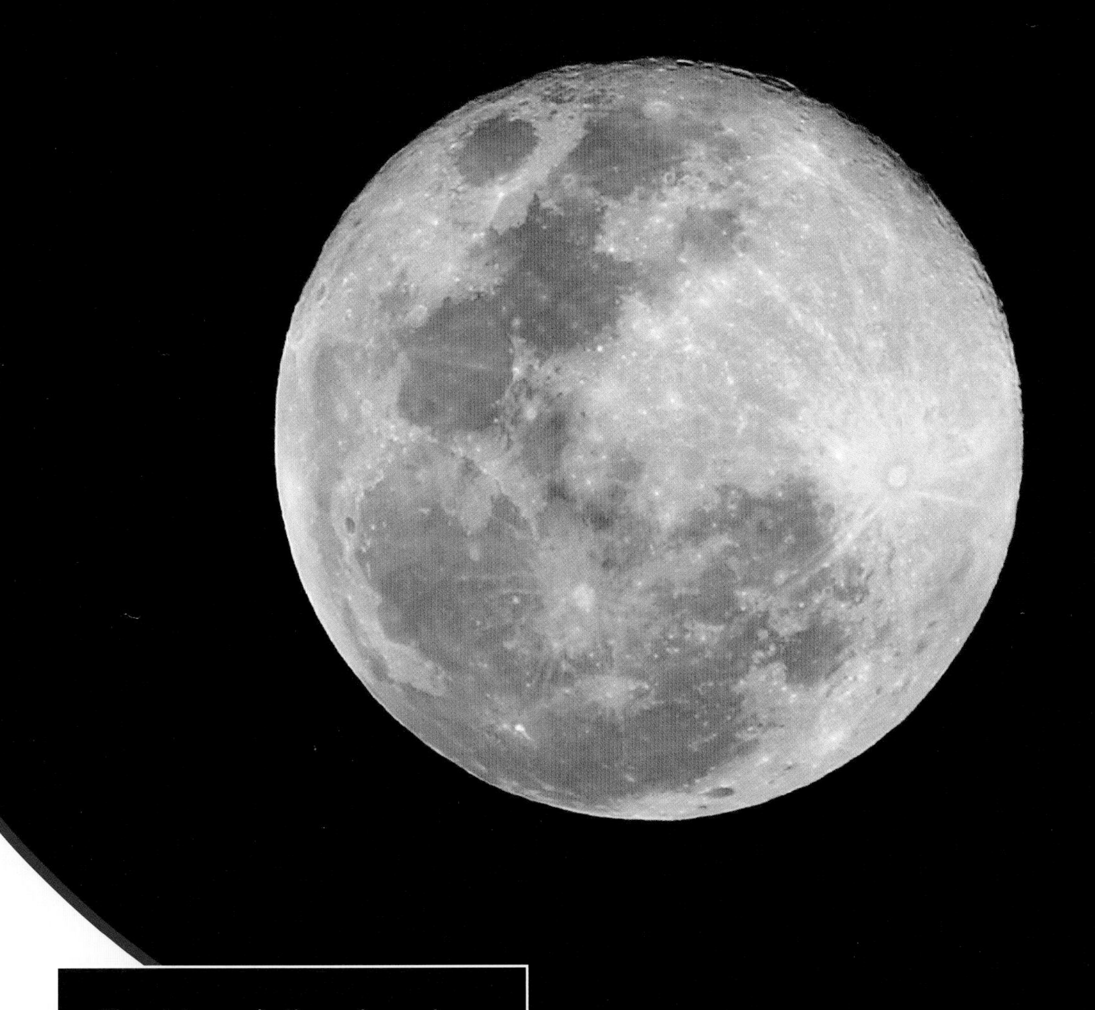

The Moon is the closest natural object to Earth in space.

The brightest light in the night sky is usually the Moon. It is smaller than stars and planets in space, but it looks large to us because it is much closer to Earth.

Day and Night

By day, we don't usually see the Moon. During the day, the brightest light in the sky is the Sun. The Sun is our nearest **star**. It gives us light and warmth.

Daylight comes from the Sun, even when the sky is cloudy.

From space, you can see where it is daytime and nighttime on Earth.

Earth spins around every 24 hours. When it is daytime, we are facing the Sun. The other side of the world is facing away from the Sun, so it is nighttime there.

Earth, Sun, and Moon

Earth moves around the Sun in a path called an **orbit**. It takes a year to travel all the way around. The Moon takes about 27 days to move around Earth.

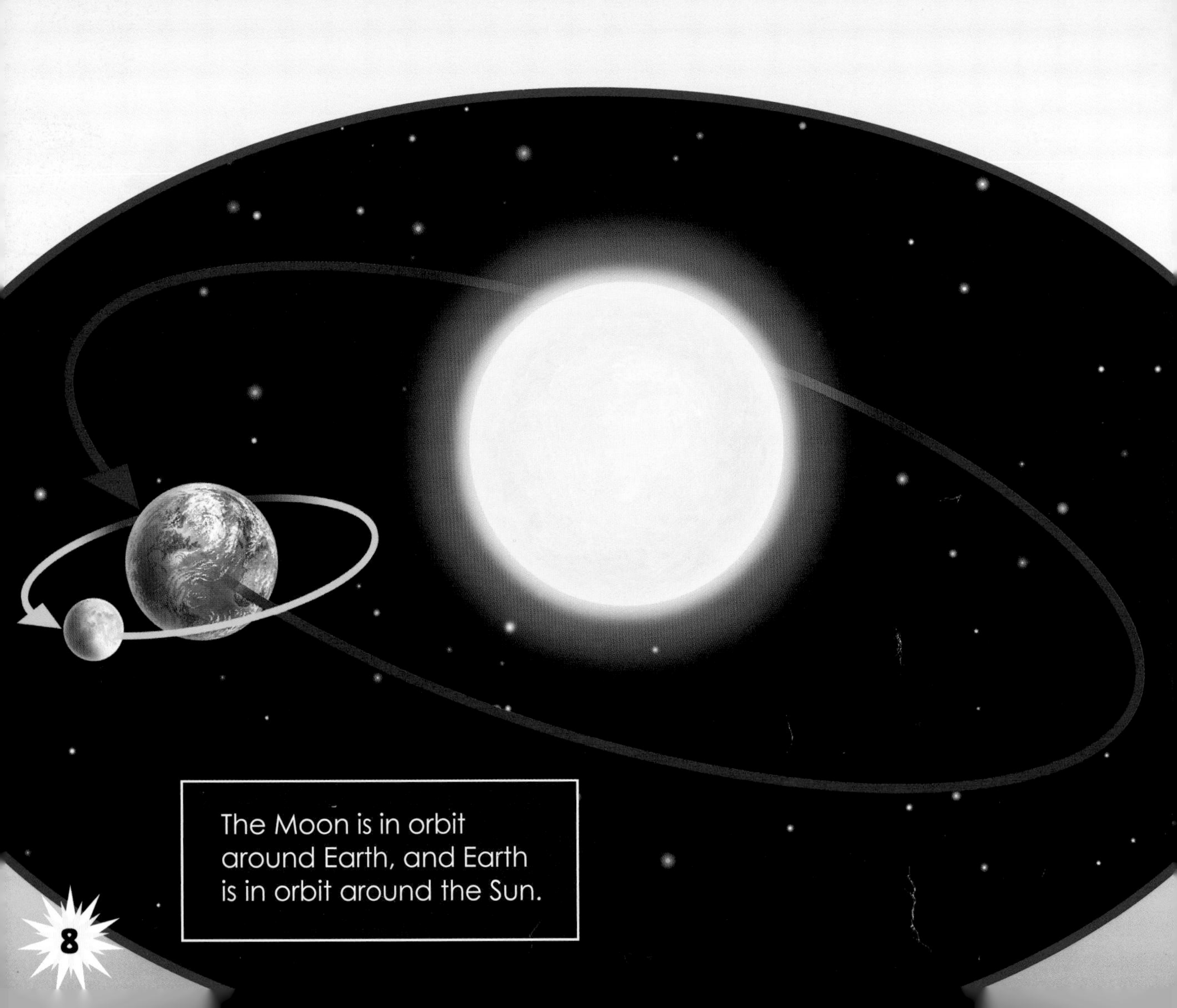

The Moon is in orbit around Earth, and Earth is in orbit around the Sun.

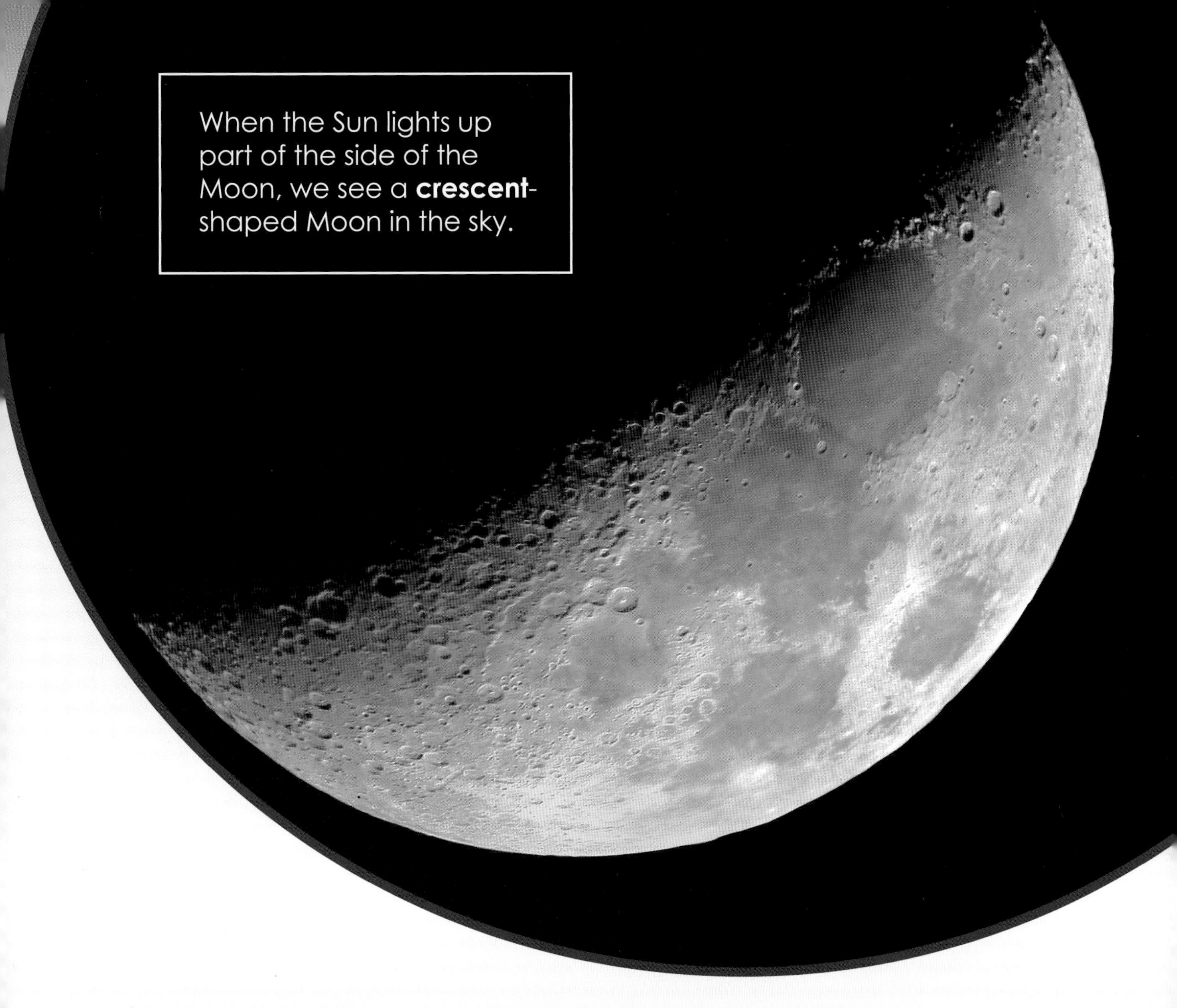

When the Sun lights up part of the side of the Moon, we see a **crescent**-shaped Moon in the sky.

The Moon does not make light itself. We see the Moon because light from the Sun shines on it. As the Moon and Earth change position, we see different parts of the Moon light up.

Solar Eclipse

Sometimes you can see an amazing sight in the daytime. When the Moon passes between Earth and the Sun, it blocks out some or all of the Sun. This is called a solar eclipse.

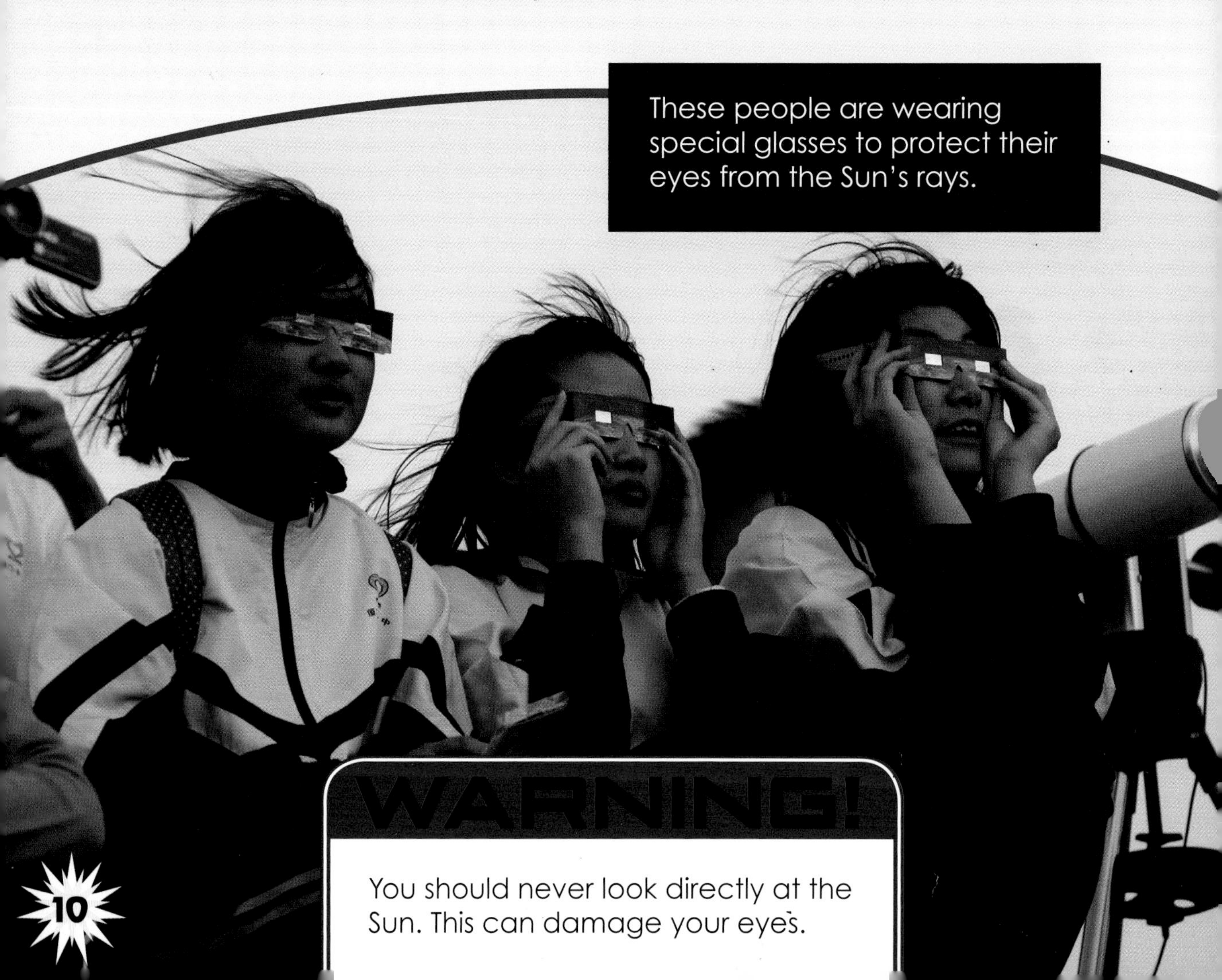

These people are wearing special glasses to protect their eyes from the Sun's rays.

WARNING!

You should never look directly at the Sun. This can damage your eyes.

Slowly, the dark disk of the Moon blocks out the Sun.

At first, the Moon starts to cross the Sun. It looks as if someone has taken a bite out of the Sun. The bite slowly gets bigger.

Total Eclipse

At last, the Sun is completely covered by the Moon. The sky goes dark. This is a **total eclipse of the Sun**. The Sun's outer layer appears as a bright circle around the Moon.

This total eclipse was seen from Kenya in Africa.

The Sun can be totally covered for several minutes, or just a few seconds.

Slowly, the Moon begins to move, uncovering the surface of the Sun. The first rays of light shine through. It looks like the sparkle of a diamond ring in the sky.

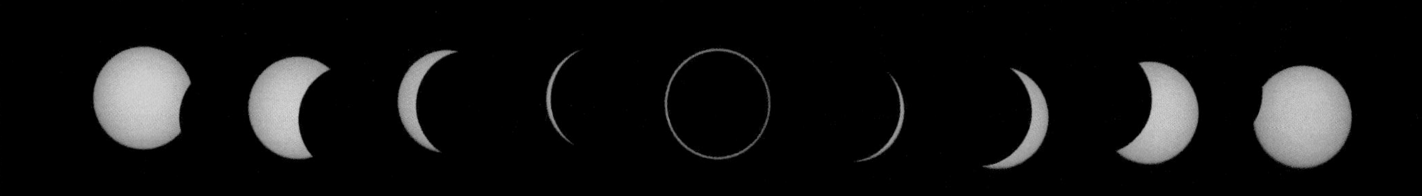

These pictures show the stages of a solar eclipse over the Pacific Ocean.

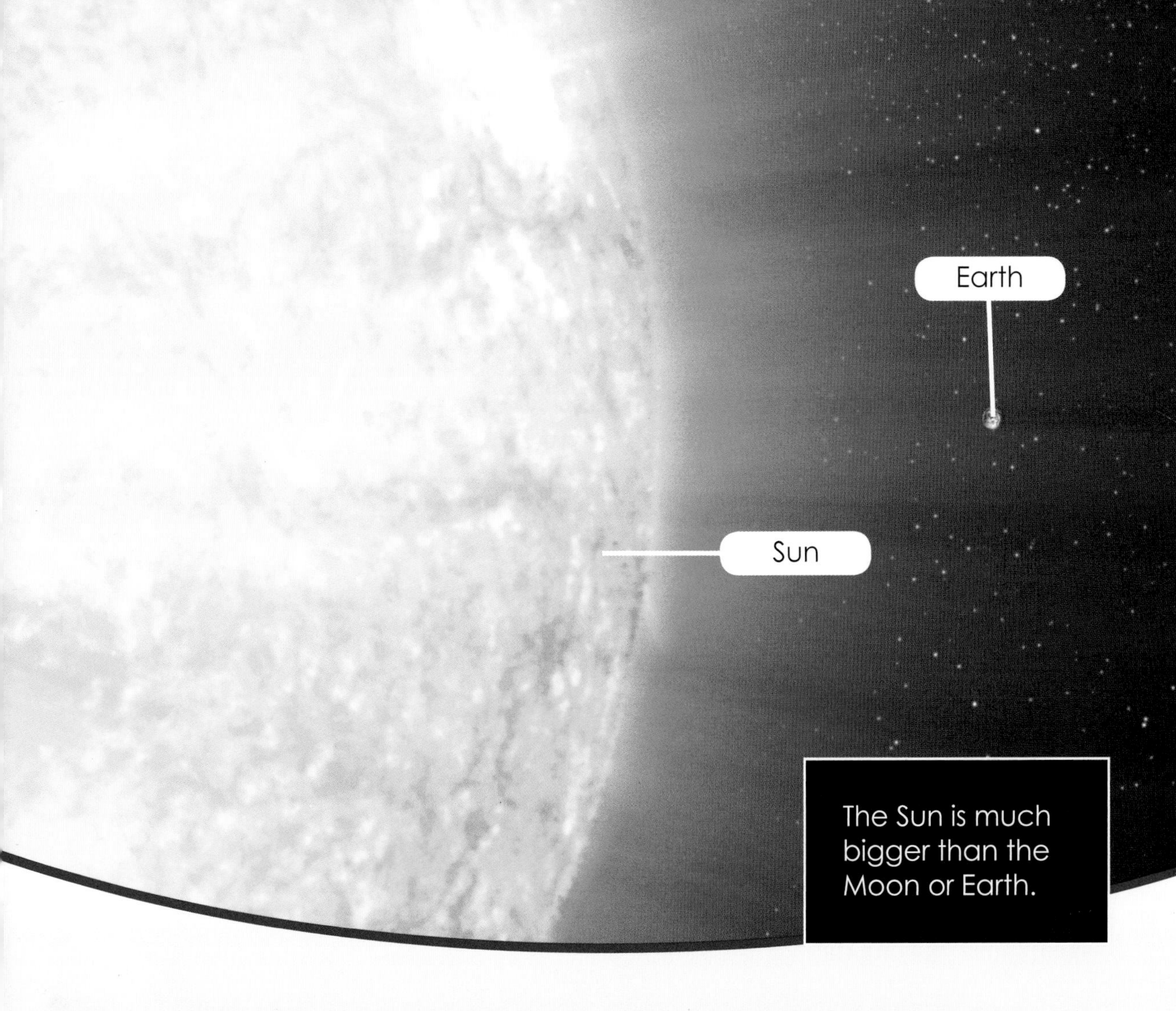

The Sun is much bigger than the Moon or Earth.

Although the Sun is around 400 times bigger than the Moon, the Moon is 400 times closer to Earth. This means that they look the same size in the sky.

Animals and Eclipses

Some animals and birds are confused by eclipses. Most animals are awake in the daytime. They think it is time to go to sleep when the sky goes dark during an eclipse.

For birds like these pelicans, an eclipse feels like the start of nighttime.

Nocturnal animals such as owls come out to hunt when the sky goes dark.

Owls and bats usually fly at night. They think it is time to wake up when the Moon blocks light from the Sun.

Eclipse of the Moon

Sometimes Earth lies between the Sun and the Moon. When this happens, the Moon passes through Earth's shadow and goes dark. This is called a **lunar eclipse**.

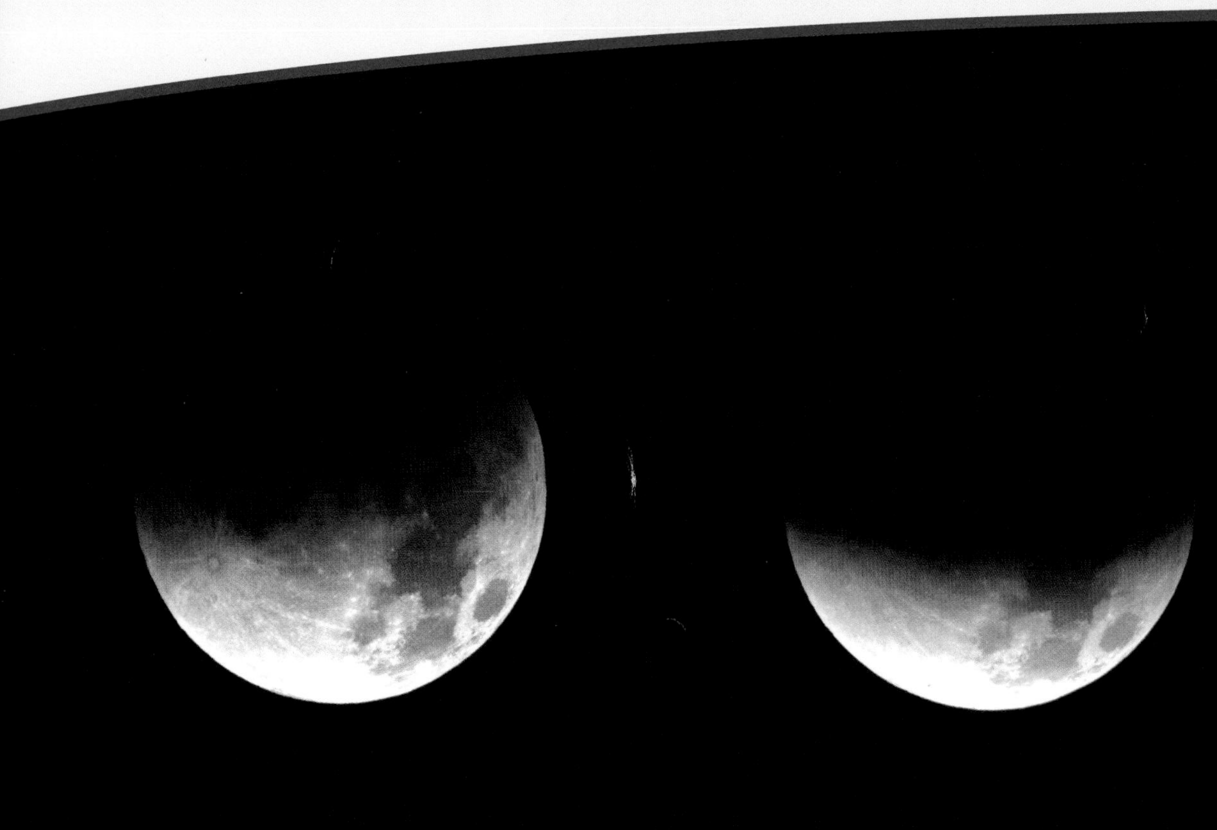

Earth stops the Sun's light from reaching the Moon. The Moon looks red in the night sky. A total lunar eclipse can last as long as an hour.

Earth's shadow slowly crosses the Moon during a lunar eclipse.

Exploring Eclipses

Astronomers are people who study the night sky. During an eclipse, they can see the outer layer of the Sun very clearly. Normally, this layer cannot be seen because the Sun is too bright.

Astronomers use special equipment to study the Sun during an eclipse.

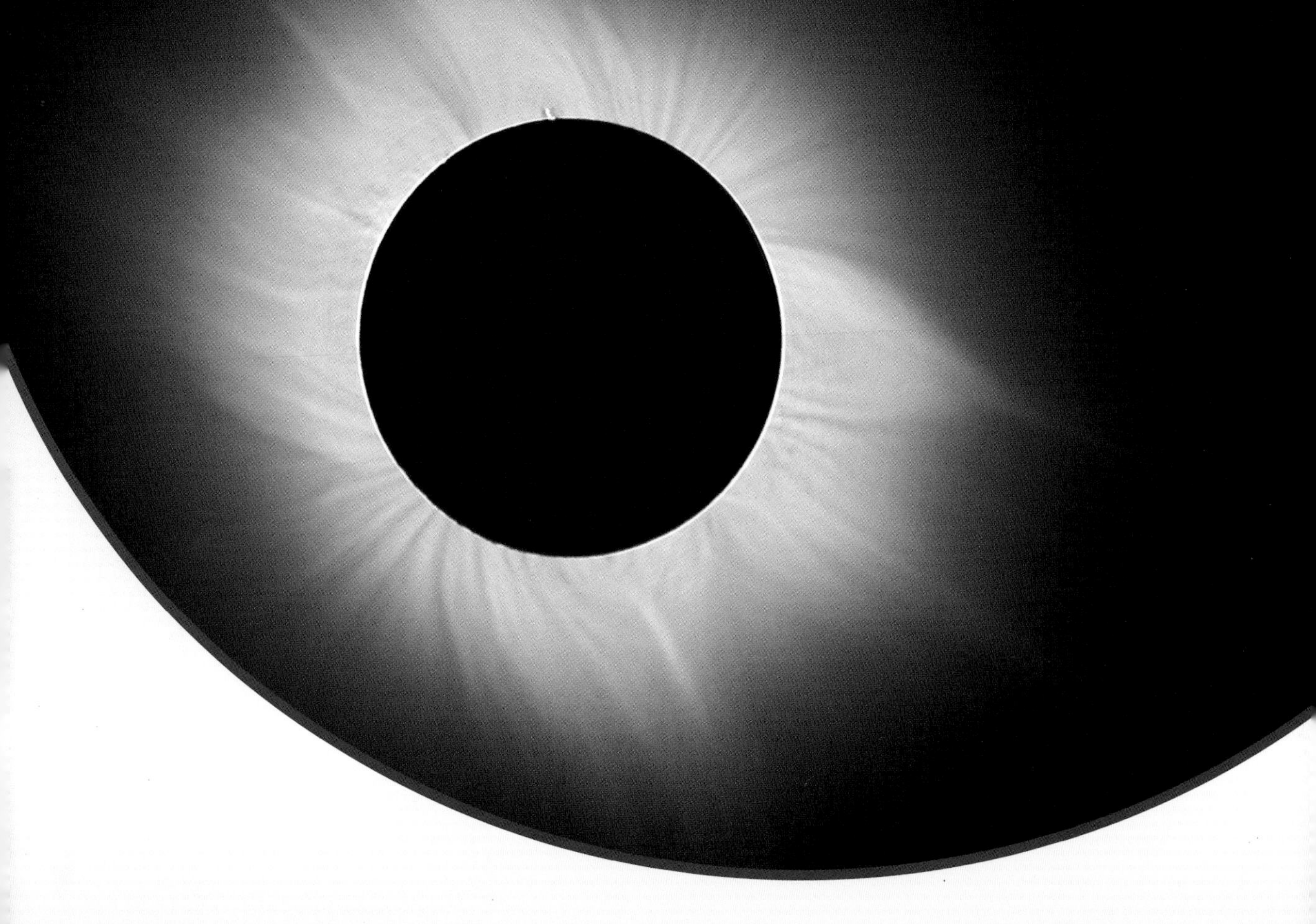

The glow around the dark disk is the Sun's corona.

This outer layer is called the Sun's **corona**. It is incredibly hot. The corona stretches millions of miles into space.

Across the Solar System

There are seven other **planets orbiting** the Sun. Some have many moons orbiting around them. There are more solar eclipses on these planets than we have on Earth.

You can see the shadows of three of Jupiter's moons on the planet's surface.

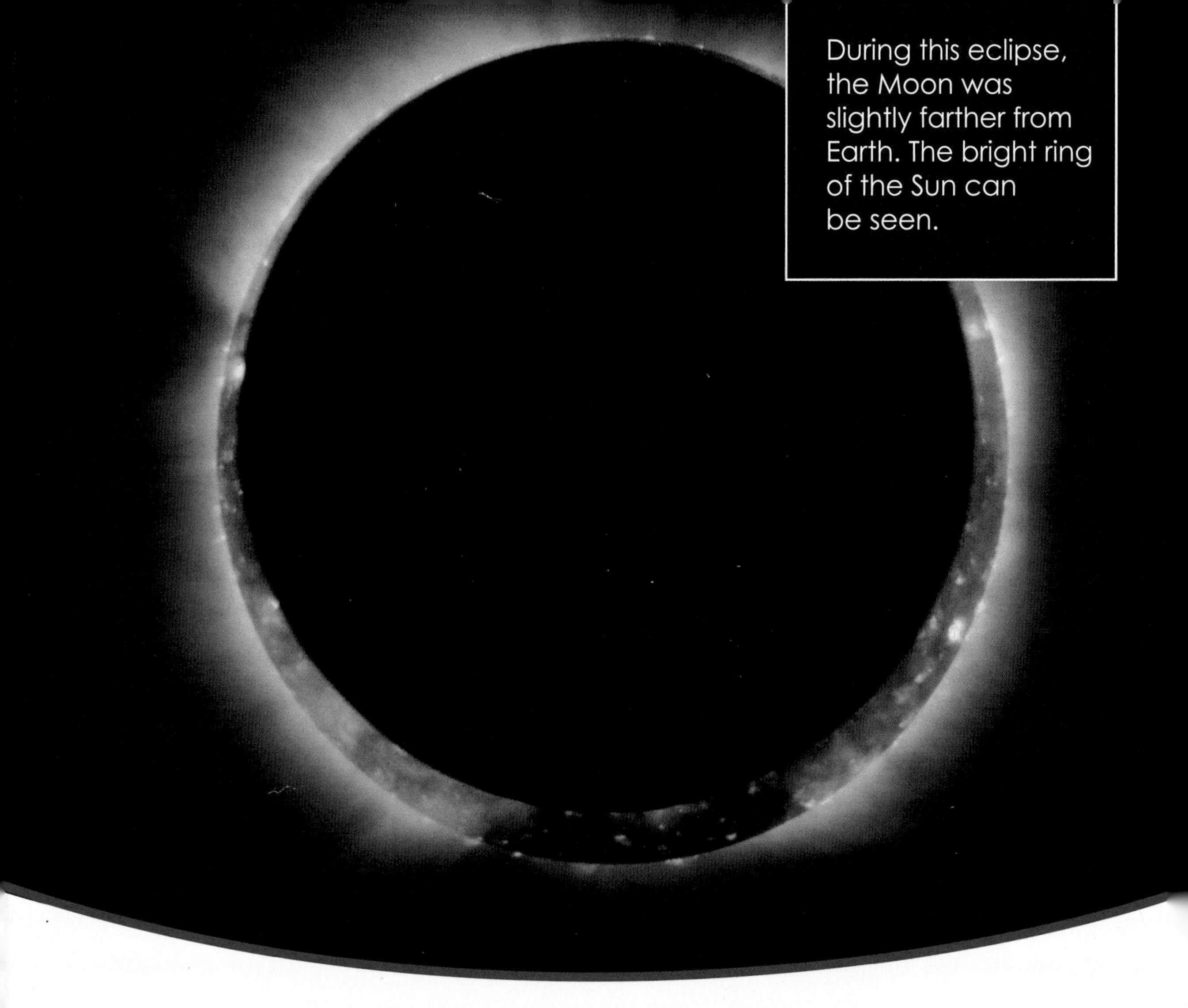

During this eclipse, the Moon was slightly farther from Earth. The bright ring of the Sun can be seen.

The Moon and the Sun usually look as though they are the same size to us on Earth. On many other planets, eclipses can cover part of the Sun or make it completely disappear.

Eclipses in the Past

Ancient people did not know why eclipses happened. Some people believed a **demon** or an animal eating the Sun caused an eclipse.

Ancient Egyptians recorded eclipses more than 4,500 years ago.

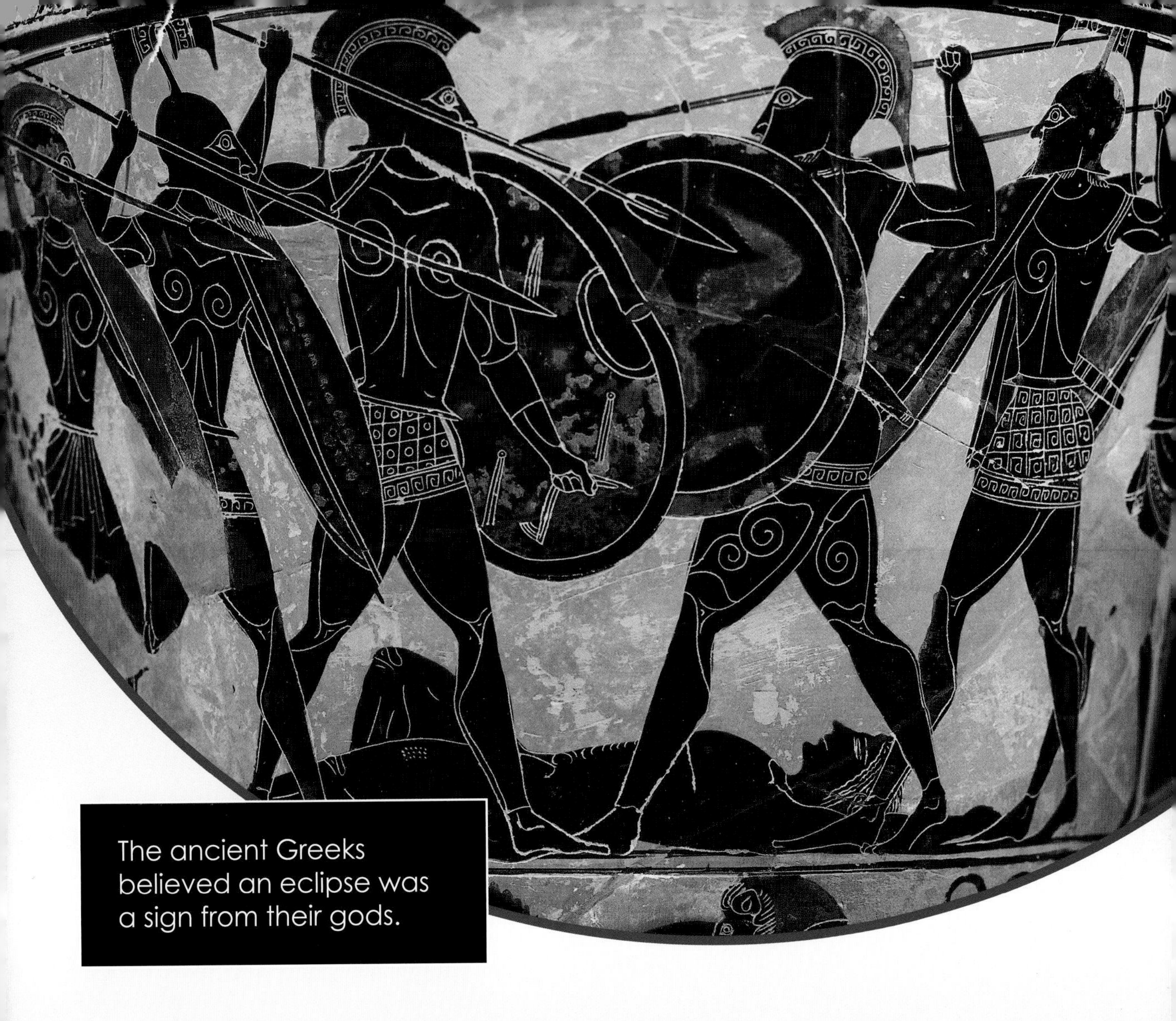

The ancient Greeks believed an eclipse was a sign from their gods.

Eclipses have changed history. Long ago, the ancient Greeks were fighting a battle against a Turkish army. When they saw an eclipse, both sides put down their weapons and stopped fighting.

See for Yourself

To see an eclipse, you must be in the right place at the right time. Some people travel around the world to watch eclipses. There will be a **total eclipse of the Sun** across North America on August 17, 2017.

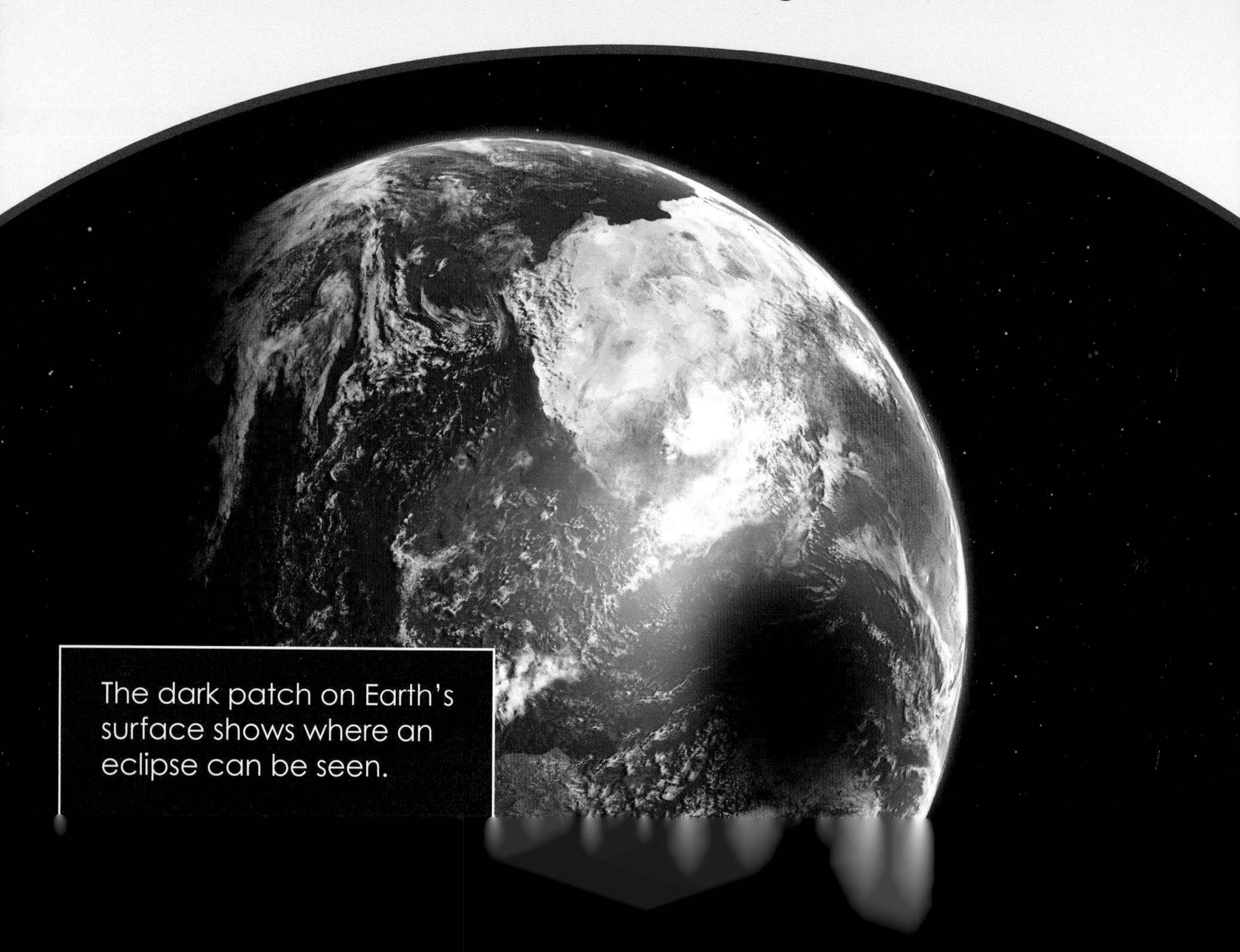

The dark patch on Earth's surface shows where an eclipse can be seen.

This telescope is being used to show an eclipse on a screen, so that it can be watched safely.

You should never look directly at the Sun. You can use special glasses or a **pinhole projector** to view an eclipse. People can also use a **telescope** to show an eclipse on a screen.

Model an Eclipse

You can show people what a solar eclipse looks like by using a flashlight to model the Sun and a ball to act as the Moon.

What you need:

- a plain wall in a dark room
- a flashlight that makes a circle of light
- string
- a ball.

What to do:

- Tie a piece of string around the ball.
- Ask a friend to shine the flashlight on the wall so it makes a circle of light.
- Hold the ball by the string.
- Slowly move the ball in front of the flashlight. The effect on the wall should look like a solar eclipse.

What does this experiment tell you about the way that light moves?

Glossary

astronomer person who studies space and the night sky
corona Sun's outer layer, which can only be seen during an eclipse
crescent curved shape made by the Moon when only part of it can be seen
demon evil spirit or monster from stories
lunar eclipse when Earth moves between the Sun and the Moon, stopping light from reaching the Moon
orbit path that an object in space takes as it moves around another object, such as when Earth goes around the Sun; also the act of moving along that path
pinhole projector simple device that can be used to view an eclipse safely
planet large object (usually made of rock or gas) that orbits a star. Our planet, Earth, goes around the Sun.
star huge ball of burning gas that produces massive amounts of heat and light
telescope device that astronomers use to make things in space look bigger
total eclipse of the Sun when the face of the Sun is completely covered by the Moon
universe everything in space, including Earth and millions of stars and planets

Find Out More

Books

Hansen, Amy S. *Where Does the Sun Go at Night? An Earth Science Mystery*. Mankato, Minn.: Capstone, 2012.

Landau, Elaine. *The Sun* (True Books: Space). Danbury, Conn.: Children's Press, 2008.

Morrison, Jessica, and Steve Goldsworthy. *Eclipses* (Space Science). New York: Weigl, 2011.

Owen, Ruth. *Solar and Lunar Eclipses* (Explore Outer Space). New York: PowerKids, 2012.

Web sites

Facthound offers a safe, fun way to find Internet sites related to this book. All of the sites on Facthound have been researched by our staff.

Here's all you do:

Visit **www.facthound.com**

Type in this code: 9781432975159

Index

ancient peoples 24–25
animals, behavior of 16–17
astronomers 20, 30

bats 17
beliefs of the past 24–25
birds 16, 17

corona 21, 30
crescent Moon 9, 30

daytime and nighttime 7, 16, 17

Earth 4, 7, 8, 9, 15, 18, 19, 23
Egyptians, ancient 24
eye protection 10, 27

glasses 10, 27
Greeks, ancient 25

Jupiter 22

light 6, 9
lunar eclipses 18–19, 30

modeling an eclipse 28–29
Moon 5, 6, 8–9, 10, 11, 12, 13, 14, 15, 18–19, 23
moons 22

orbits 8, 22, 30
owls 17

pinhole projectors 27, 30
planets 4, 22–23, 30

solar eclipses 10–15, 22, 23, 28, 29
stars 4, 6, 30
Sun 6, 7, 8, 9, 11, 12, 13, 14, 15, 18, 19, 20–21, 23

total eclipses 12–13, 19, 26, 30

universe 4, 30

watching an eclipse 10, 26–27